William Larder

Thirty Years at the Cutting-Board

Being a work designed to assist the student to acquire knowledge in the

art of cutting. Containing a series of diagrams laid down to measure

William Larder

Thirty Years at the Cutting-Board
Being a work designed to assist the student to acquire knowledge in the art of cutting. Containing a series of diagrams laid down to measure

ISBN/EAN: 9783337256890

Printed in Europe, USA, Canada, Australia, Japan

Cover: Foto ©berggeist007 / pixelio.de

More available books at **www.hansebooks.com**

THIRTY YEARS AT THE CUTTING-BOARD:

BEING A WORK DESIGNED TO ASSIST THE STUDENT TO ACQUIRE KNOWLEDGE IN

THE ART OF CUTTING.

CONTAINING

A SERIES OF DIAGRAMS LAID DOWN TO MEASURE.

AND AS USED BY US IN OUR DAILY PRACTICE

BY

WILLIAM LARDER.

INTRODUCTION.

Having been frequently requested to give instructions in cutting, but not having the time to devote to that subject, we have concluded to put our rule in the simplest possible form. Having used it in its present form without any material change for the past ten years, we believe we have succeeded in putting it in a form so that any tailor of ordinary intelligence can be his own teacher, and will receive instructions the value of which cannot be estimated. The rule is the product of thirty years' experience and continued practice. Our design is to instruct the uninitiated. All those whose present methods of cutting prove unsatisfactory to them, we would advise to try our system; although only designed to instruct those who are about to adopt cutting as a profession, we believe the simplicity of this rule will recommend it to many, being well adapted for drafting garments on the cloth.

GENERAL REMARKS ON GOODS.

CUTTING AND MAKING.

It requires good judgment and long practice to estimate the proper size to cut garments from goods of different texture, some goods being elastic, while others are stiff and hard, and will not give in making motions with the arms. The elastic goods will have to be cut close to or inside the measures; hard, stiff goods must be cut larger than the measure; very stiff goods we usually cut two sizes larger in widths, or the garments will look and feel too small. This rule must be observed in coats and pants; it is not necessary to make any difference in vests, for the interlining in vests is always of stiff goods; therefore vests, cut from any material, will be the same size when finished. Practice only will give the student the idea as to how much difference to make for the different kinds of material. No very definite instructions can be given on this head; therefore we must leave it with the student, for by practice only can the difficulty be overcome.

OVERCOATS.

For overcoats we have to make sufficient allowance to go over the undercoat. The difference we make is to allow 2 inches on the breast measure, if *the measure is taken on the vest*, more or less according to the material used. But in most cases we observe the following rules: For a surtout overcoat we add ¾ of an inch to the first measure down the back. To the short blade, ½ in.; to the length of shoulder, ½ in.; and to the measure over the shoulder, ¾ in.; ½ an inch to width of back; ½ an inch to the length of waist, and ½ to ¾ of an inch on the length of sleeve; 2 inches to the breast measure, 2 on the waist, 2 on the hips, and 2 on the seat when all the measures are taken on the vest and pants; in all other respects drafted the same as Diag. A, except to drop the square ¾ of an inch below A, to get point K. Whatever we allow on the first measure down the back we always drop the square the amount added to the measure. Diagram A should be thoroughly learned before making an attempt to draft the other diagrams, for reference will often be made to that diagram.

SACK OVERCOATS.

On the first measure down the back, we allow in most cases 1 inch; 1 inch to the shoulder length, 1 inch to the measure over the shoulder to the first measure down the back, and 1 inch to the blade. When drafting a sack-coat add ½ an inch in front of the construction line for a normal figure, and increase the distance as the waist enlarges. We add 1 inch when the waist is 2 inches smaller than the breast, and for every 2 inches that the waist increases we add ½ an inch, so that when the breast and waist measures are equal, we add 1½ inches in front of the construction line. We observe the same rule for all sacks that are cut to button all the way down. When cutting a coat for a stooping figure, observe the diagrams for these forms; the changes have been there described, only the back line that is carried forward at A, to produce the round, will have to be changed more or less, according to the degree of roundness of back. When you meet with a person out of the regular shape, always take care to note it. If one shoulder should be lower than the other, put the square under each arm, and note the difference by a mark on the back, at the centre, above or below the first mark made, as the case may be, and when cutting a coat for a form of this kind, take the shoulder off ½ an inch at the top, and make other changes as shown on plate 20.

Sometimes, men will be met with, with one very prominent shoulder-blade, and the other very flat ; be careful and note which of the two sides is the largest so as not to make a mistake when cutting, for the small side will have to be cut flat for such a form, and the other side piece will have to be cut with more round and perhaps a little wadding put in the flat side ; but it is simply impossible to note all the difficulties to be encountered, therefore we will only advise great care in observing the different builds, and study what changes to make for the various forms.

MAKING.

When making a frock-coat, we desire to have the following rules observed : First, always cut the canvas to correspond with the forepart ; if any cuts in the forepart cut the canvas also. Press the round on the front of forepart back, never to the front, for by so doing the lapel would be too round on the front, and would have to be drawn in with the edge-stay and make it very hard to get in the proper shape. Sew the back a little easy on the round of side piece, and when joining the shoulder-seam, let the back go on easy. If the shoulder at the forepart should chance to be cut too long, let it run out at the seye, and then make the shoulder have the proper curve. Never mark the piece off too sudden ; perhaps the best way would be to put a thread in and make an outlet of the surplus. Never put the linings in small; have plenty of length and width in the shoulder. In making up the skirt press the round on the plait over the hip ; press it until straight, and baste it on the back easy, that is, not too short ; if put on too short, it causes the plait to roll open, and keeps the upper back from lying close to the under back, giving it the appearance of having too little spring. Put a thread around the seye from a point, say 2 inches in front of the shoulder point, and around the under part of seye to a point ½ an inch above the notch for the front of sleeve, and then stretch it between these two points. Cut the canvas and lining in the centre of the part stretched and put in a V. In making the collar, after it is padded put a strong thread in the crease line and draw it in well ; this is a better way than stretching the fall and stand edges, to give it the appearance of being drawn in on the crease. When putting on the collar it should be sewed on a little snug in front and just fair the remainder of the way. Be careful not to put it on too short ; if too short it will cause the coat to crease in front of seye and set in too close at the waist, and give the coat the appearance of being too long in the back, and ride up in the neck ; if put on too long it will cause the coat to swing out at the waist when unbuttoned, and when buttoned, the collar will set out from the neck. The same difficulties will occur if point K is carried too far forward or too far back, or if the shoulder be too short or too long. If too short, it has the same appearance as a short collar. Point K being too forward has also the same appearance as a short collar. If the shoulder is too long or point K too far back, it has the same appearance as a long collar. In making up cutaway coats, be careful not to put the edge-stay on too tight ; if too tight it will cause the coat to break back of the buttons and button-holes. The stay should be put on just tight enough to prevent the edge from stretching in pressing. Be careful and have the facing put in so that it does not roll to the front. Sometimes the canvas rolls forward when the facing does not ; this is caused by the canvas being too long at the back edge.

SACK-COATS.

Sack-coats should have the back put on a little long on the blade, commencing about five inches from the top of side point; if on too short at this point it will give the forepart the appearance of being too high at the side point and cause it to crease, the creases running under the arm; it will also throw the coat out at the bottom, and make it look as if it had too much spring. Never draw the front edge in much; just put the stay on the least snug to prevent the edge from stretching. Except in large sizes, when the front is cut round on the front, then draw it in until it is nearly or quite straight. We refer now to sacks that are cut to button all the way; oversacks to be treated the same way. We prefer to have the linings of all sack-coats left open at the bottom.

VESTS.

Rules to be observed when making vests: Be careful never to stretch the front edge; if with a collar, put the collar on a little snug to prevent it flaring out on the crease when it is buttoned. If without a collar, put a stay on the edge from the back of neck to the top button; put it on a little tight, for this part being on the bias is easily stretched. Stretch the front of shoulder and the shoulder seam, and put the interlining and lining in a little full; this will prevent the vest from creasing on the shoulders.

The directions for pantaloons are given with explanation of the diagram.

THE AUTHOR.

DIRECTIONS FOR TAKING MEASURES.

This being the most essential thing to acquire in order to be successful, great care must be exercised to have the measures properly taken, and for a new beginner it is one of the most difficult things to master; in fact, it may be asserted that the most experienced very often get mistaken. We have had thirty years' experience and cannot claim perfection. But we claim that our system of measurement is all that is required to produce a well-balanced garment when applied to the rule as laid down. We conceived the idea, several years ago, of putting a spirit-level in a common square, a diagram of which is shown on Plate 1. We find it to answer our purposes well. We prefer to take all measures for coats on the vest, believing that we can get them more correct in this way. When the coat is taken off, adjust the vest properly, seeing that the back seam is in the centre of the back; this being done, place the square under the arm, pressing it up tight, to be sure that nothing prevents it from going in the proper place; after this has been ascertained, let it drop until it just touches without pressure. Now see that the square is on a level, and then make a mark on top of the square front and back of the arm. We then turn the square with the short side up, and press it close to the body and front of shoulder, and mark inside the square. We then turn to the back, and mark a point in the centre even with the bottom of the seye. Now take the inch measure, place the end on the socket-bone, and call off the measure to the first mark on the back, then to the natural waist, full length of waist, and length of coat; next measure across the back to get the width, continue down to the elbow and wrist for length of sleeve, then around the seye; take the length of shoulder from the top of back at socket-bone to the bottom of seye in front; then the short blade measure from the upright mark in front of seye to the centre of the back; measure for the height of shoulder from the bottom of the seye in front to the first mark on the back. Measures can also be taken from the socket-bone around the front of shoulder, and under the arm to the centre of the back; this we call the long blade measure. Continue the measure up to the starting-point for the upper shoulder measure; these two measures we take to test the correctness of the short measures. Now measure around the chest close under the arms, around the waist, hip and seat; these measures should be taken snug, but not tight—in fact, none of the measures should be taken tight; on the long blade and upper shoulder we allow one inch on each; if the blade measure is taken too loose, it will throw point K too far forward; if taken too tight, point K will be too far back.

TO MEASURE FOR A VEST.

First from the socket-bone down to the place for the top button, and continue to the full length in front; also measure to the top of the hip-bone. This measure will give the place for the bottom of the back strap. When a measure is taken for a double-breasted vest, measure for the opening in front, for the lengths as before stated.

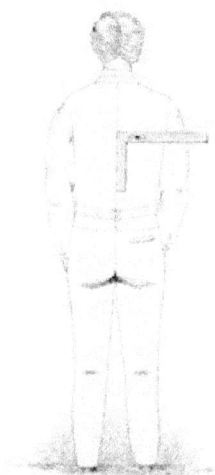

TO MEASURE FOR PANTALOONS.

First measure from the top of the hip down the outside to the seam of the heel of the boot or shoe, take the inside seam close up in the crutch to the heel as before stated. Measure around the waist, seat, thigh, knee and bottom, all to be carefully taken.

We will again urge upon the new beginner the necessity of having correct measures, and a continual practice of drafting to measure; by persevering in the study you will gain confidence. Take all the measures you can of your friends, *and more than once*, and then compare the measures first taken, and see how they correspond with the last taken; by so doing you will become more proficient in taking measures. We don't insist that our way of taking measures is the only correct way, any way will do, providing the measures are correct; but bear in mind that the length of shoulder, the proper length of back to the bottom of the seye, height of shoulder, and the blade measures must be correct to have a well-balanced coat.

On the plate facing page 8 will be found three figures showing how to place the square, the instructions in regard to which have been previously given.

PLATE 1.

DIAGRAM A.

TO FORM THE DRAFT.

First draw a line down the edge of the paper or cloth. Then move the square in at F 1½ inches, and draw a second line up to A. Then mark from A to B ¼ to C ⅓, to D ½ of the breast measure. To E 1 inch less than the breast measure, say 17 inches. Full length of waist to fancy or fashion. Square across from the inside line at A, B, C, D, E, and F. Mark from A to Y ¼, from C to G ¾ and 1¼ inches. Mark across from B to C at G, draw a line from X to E. The width of back at the waist 2½ inches or what the fashion may be. Form the remainder of the back as shown in the diagram, going above the line at Y ½ an inch and as much as ¾ for a 48 size. Mark point H ½ the breast measure from D, or ⅔ of the short blade measure. Square down from H to S, D to I ⅔. Square up to P, I to J ½ less ¾ of an inch. Put the corner of the square on point J, with the long arm resting on A, and square up to K for the neck point. Measure from a point 1 inch in front of I up to W ⅓ of the breast measure and ¾ of an inch more; D to M, half of the full breast measure and 2½ inches more. Square up from M and over to P. M to N ½ the shoulder length, 6⅛, being ½ of 12¼, the shoulder length; from the socket-bone, N to O $\frac{5}{16}$. Curve the side piece from X to H. Mark by the back from X to D for the upper part of side piece. Now move the back, as shown by the dotted lines, opening at X one inch. Mark a point at L ½ an inch from the dotted line of back. Form the remainder of the side piece, by curving as represented, going out to the plait line at Z. Square over from Z to S. Mark up to T 1 inch from S, and form the bottom of the side piece. Hollow the side piece and forepart ¼ of an inch each side of the line H, S, springing out at the bottom ¼ of an inch over the line. Now turn the back over to the shoulder, as represented, with the corner touching at K. Drop it ½ an inch at P, opening the shoulder point from the back ¾ of an inch. Then form the top of shoulder and scye. Draw a line from where the back touches the forepart at the line below P over to O for the front of gorge, curve the gorge, as shown in diagram, place the measure ½ an inch back of K, and sweep from Z for length of front. Measure over to V the waist measure, and 2¼ inches more to allow for seams, if the measure is taken snug. The notch in front of scye is 1¼ inches above point I, to locate the place for front of sleeve. Measure around from the front notch to the notch in the back, ½ the scye measure, the place for the back seam of sleeve.

This diagram was made from the following measures: First, down the back to the bottom of scye, 8¾; waist, 17; full length of waist, 18½; width of back, 7¼; elbow, 20½; to the wrist, 31¾; scye, 17¼. Shoulder length, 12¼; over the shoulder to U, 17¾; short blade, 11⅛; breast, 36; waist, 32; hip, 34; and seat, 38. When making a draft to measure add ¼ of an inch for seam to the first measure, ½ an inch to the blade measure, 2½ inches to the breast, and 2¼ to the waist measure, to allow for seams, etc. When the pattern is finished according to the above measures, the long blade measure will be 24, and the upper shoulder 28 inches. Hollow centre seam of back slightly, as shown.

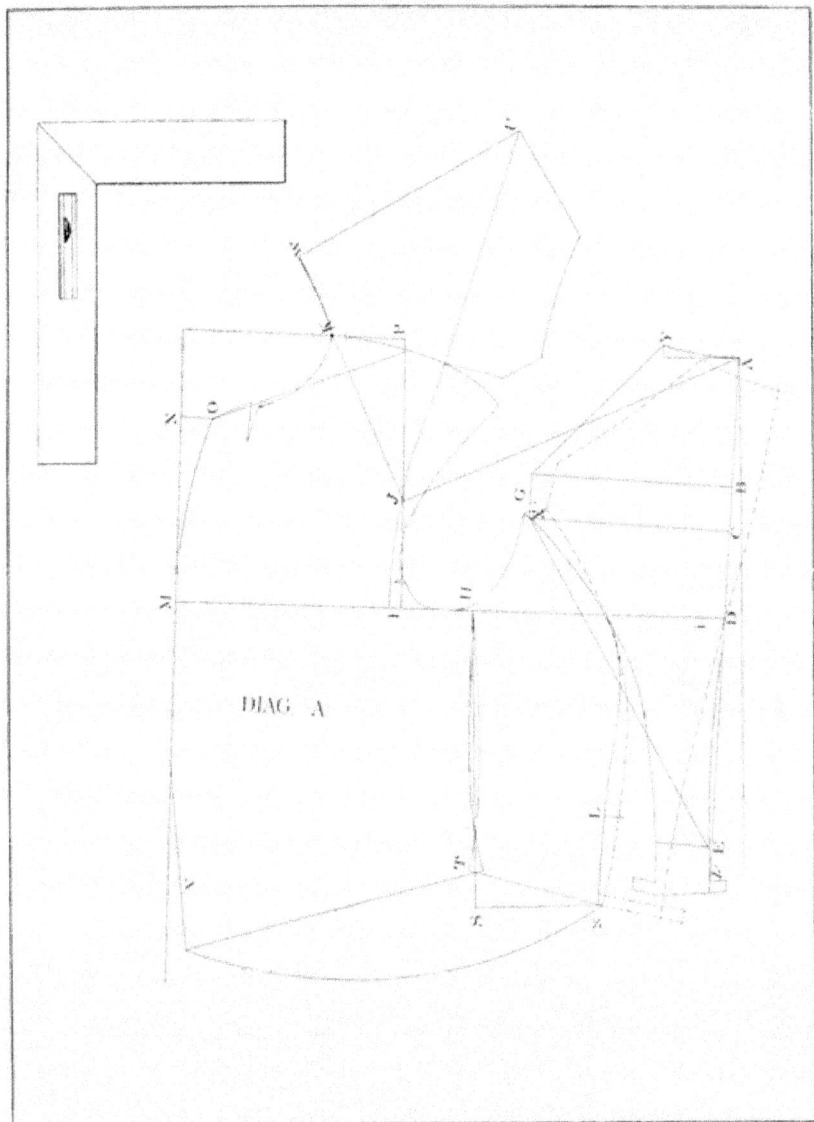

PLATE 1

DIAG A

PLATE 2

DIAG. B

PLATE 2.

DIAGRAM B

Is for a stooping figure. Drafted to the following measures: From A to D 9½, to E 18, to F 19½ ; width of back, 7¾ ; short blade, 12 ; shoulder length, 12¼ ; over the shoulder from a point 1 inch in front of 1 to U, 18 ; seye measure, 17½ ; breast, 36 ; waist, 34¾ ; hip, 39½ ; and seat, 38. Notice the difference in the draft for the back. The line at A is carried forward ⅝ of an inch. A, B, and C are squared across from the inside line, in order to shorten the distance from A to H, and to give round to the back from A to D, and allows the back to fall in at the waist. The opening at X is 1½ inches instead of one inch, as in diagram A. This is done in order to take the side piece in more at the bottom, giving less spring and more round over the blade. The opening of the lower shoulder point from the back is 1 inch in place of ¾ of an inch, as in diagram A, also one inch back of K in place of ½ an inch as in diagram A, to sweep for the length in front. From N to O is ⅛ in place of ¹⁄₁₆. The dotted line at the bottom of side piece shows the difference to be made from a proportionate to a stooping figure.

11

PLATE 3.

DIAGRAM C.

Showing how to form the skirt. First draw line A, straight with the front of fore-part, as shown in diagram. Square over to B, B to C 2¼ inches. Draw a line for the back part of skirt as shown in diagram, first laying the bottom of forepart and side piece in a line. If the seat should be larger than the regular proportion, the top of side piece must be carried forward; if smaller, then the side piece will have to be brought back to reduce the spring. If the skirt is cut to measure, measure the size of the seat on line D, E, allow-ing 2½ inches over the measure, including the width of back. Make the skirt the same length back and front; round the back of skirt ½ an inch from the straight line at E, as shown in diagram. Observe the dotted line on the forepart, at the front of gorge; this is to be added when it is required to have the coat button 4 buttons, and is done to prevent a drag on the lapel when the coat is buttoned.

DIAGRAM D.

Showing how to draft a sleeve, which should always be drafted by the seye measure. First draw line A, K, and square from O to G, O to A ¾, A to B ¾. Square across at A and B, A to C ½ the seye measure. Square down from C to L, and up to G; O to E ¼ of an inch less than ½ of O G, O to D ½ of O G; J is ¼ from C; A to I 1½ inches; draw a line from I to J; H is ¾ below the line B C. Now form the upper part by going above the line at E, and curving to the first line below D, continuing to C; form the underpart as shown in diagram. Mark up from L to P 1½ inches. The width of sleeve at the wrist to fashion or fancy.

PLATE 3

DIAG. C

DIAG. D

·

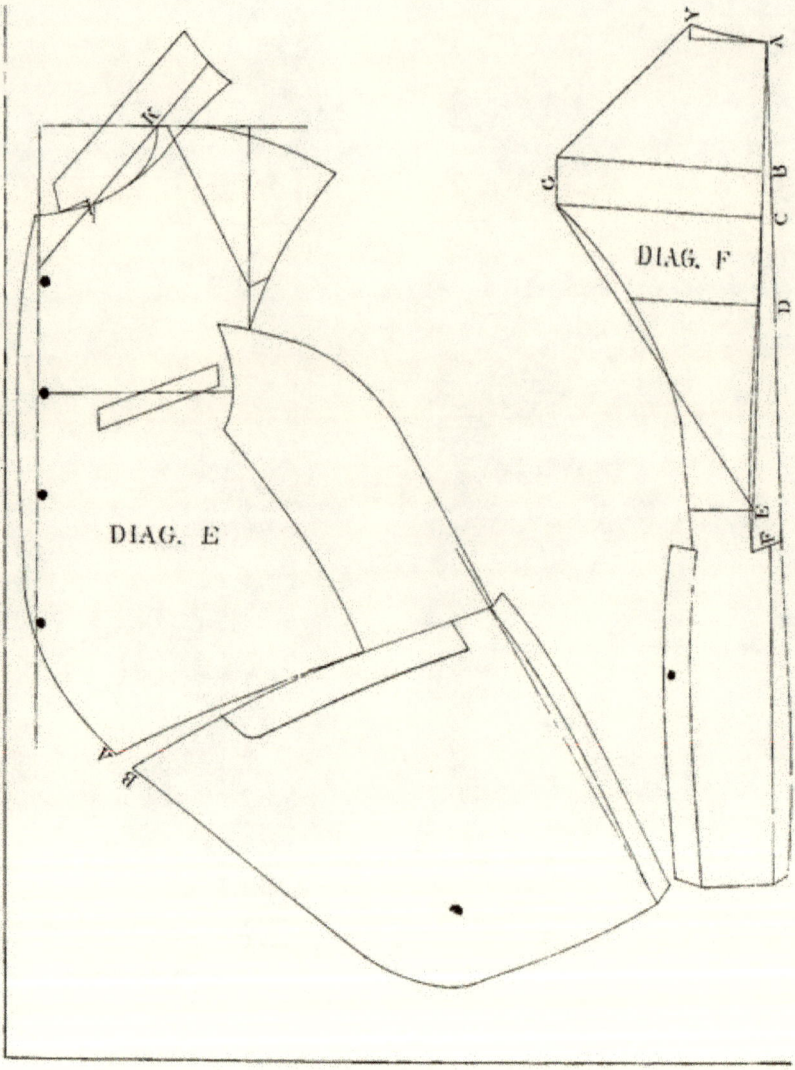

DIAG. E

DIAG. F

PLATE 4.

DIAGRAM E

Is a four button cutaway, drafted the same as Diagram A, except ½ of an inch less on the breast line at M, and ⅜ of an inch forward for the neck point at K, in order to make the collar fit close to the neck when the coat is buttoned. For the button stand add ¾ of an inch. In forming the bottom of the forepart, add ½ an inch to the waist measure to give ease and to prevent a draw on the bottom button. Place the forepart and side piece as represented, and then draw a line for the spring of skirt as shown in the diagram. Drop the skirt in front ¹⁄₁₂ of the breast measure from A to B, and then finish the skirt as represented in diagram.

DIAGRAM F.

Showing the back drafted separate. First draw a line even with the edge of cloth or paper. Move the square in at F 1½ inches, and draw a second line up to A. Square across at A, B, C, D, E and F. A to Y ½, A to B ½, to C ⅜, to D ⅜ of the breast measure. A to E 1 inch less than ½ of the full breast measure, or if cut to measure whatever the length may be, width of back from C to G ½ and 1¾ inches. Mark across from B to C at G, draw a line from G to E, width of back at the waist, 2½ inches or what the fashion may be. Finish the back as shown in the diagram.

13

PLATE 5.

DIAGRAM G.

Same as Diagram E, except being a gradual cutaway.

PLATE 5

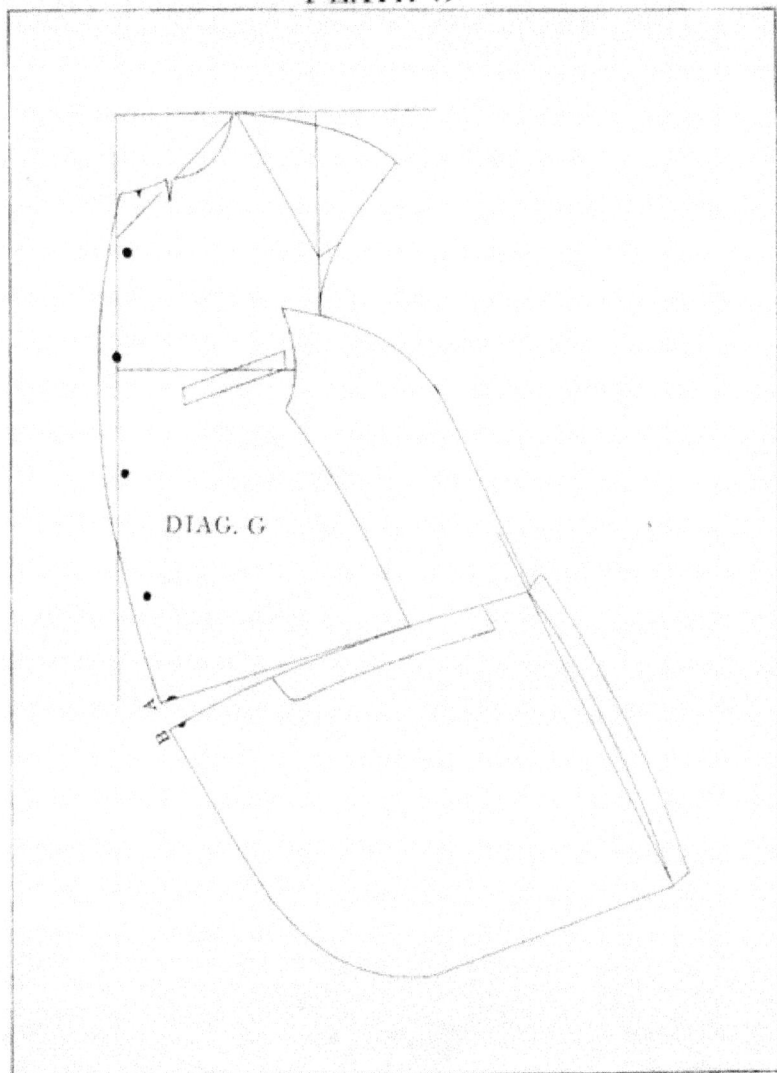

DIAG. G

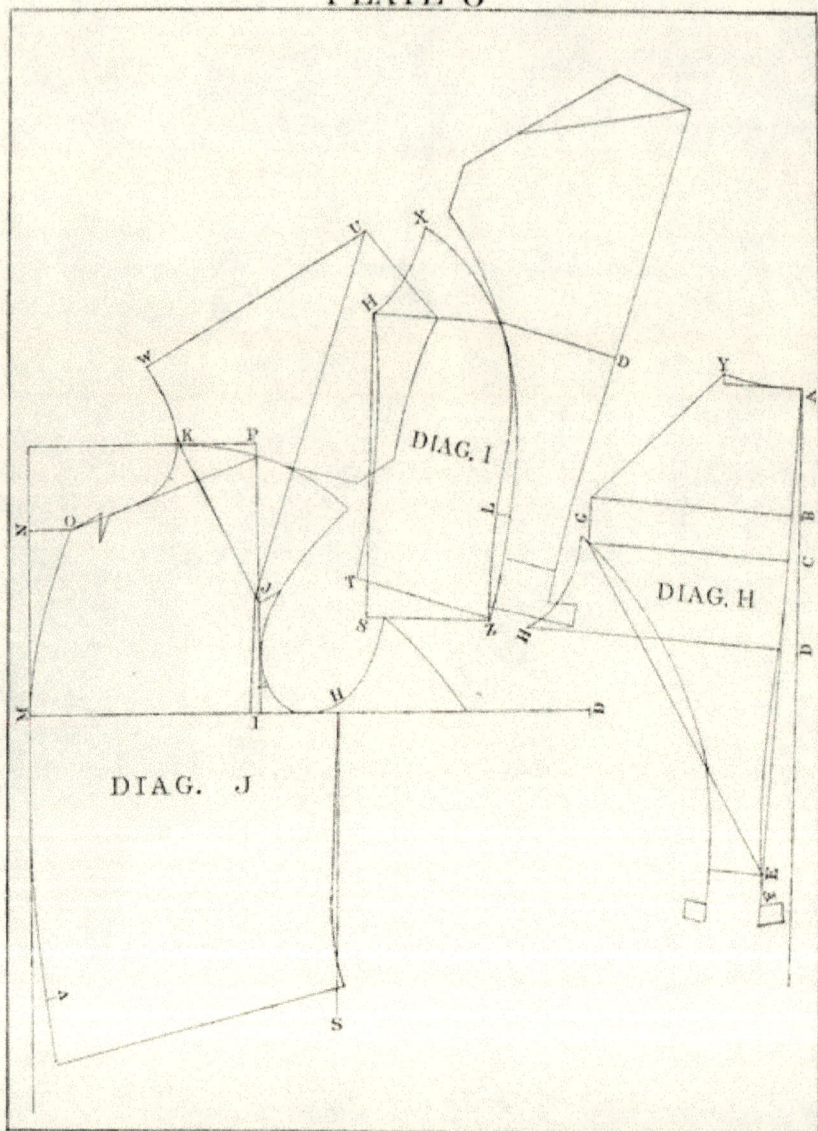

PLATE 6

DIAG. I

DIAG. H

DIAG. J

PLATE 6.

DIAGRAMS H. I and J.

Showing how each piece can be cut separate and laid down to measure. Diagram H, the back. First draw the lines as in Diagram F, then measure from A to D 8¾ and the seam. Continue down to E, 17; to F, 18½; B is ¼ of A D; C is ⅞ of A D; C to G, 7½; A to Y, ⅛ of the breast measure, and carried above the square line ½ an inch. Width of back at waist, 2¼ inches. Now shape the back as shown. Form the upper part of side piece curving to H, and by the back to line D. Square down from H to S, as shown in Diagram I. Now move the back, opening it from the side point 1 inch at X, then make a mark ½ an inch from the back at L, and curve to Z, as explained in Diagram A. Square over from Z to S; mark up from S to T 1 inch; draw a line from T to Z for the bottom of side piece. Curve the side piece from H to T as shown. This being done, bring the side piece and back down on line D M, to form the forepart. D to I, the blade measure with ½ an inch added; the remainder of the forepart to be finished as explained in Diagram A.

PLATE 7.

DIAGRAM K.

Showing the changes to be made for a dress-coat. The coat should be cut a little close to or inside the measure, usually being made to fit snug. In all other respects drafted the same as Diagram A, except where the changes are noted as in diagram. The opening of the skirt from the forepart, from A to B, is $\frac{1}{12}$ of the breast measure; the width of skirt to fancy or fashion. Cut the forepart about $\frac{1}{2}$ an inch longer in front than you would for a frock-coat.

PLATE 7

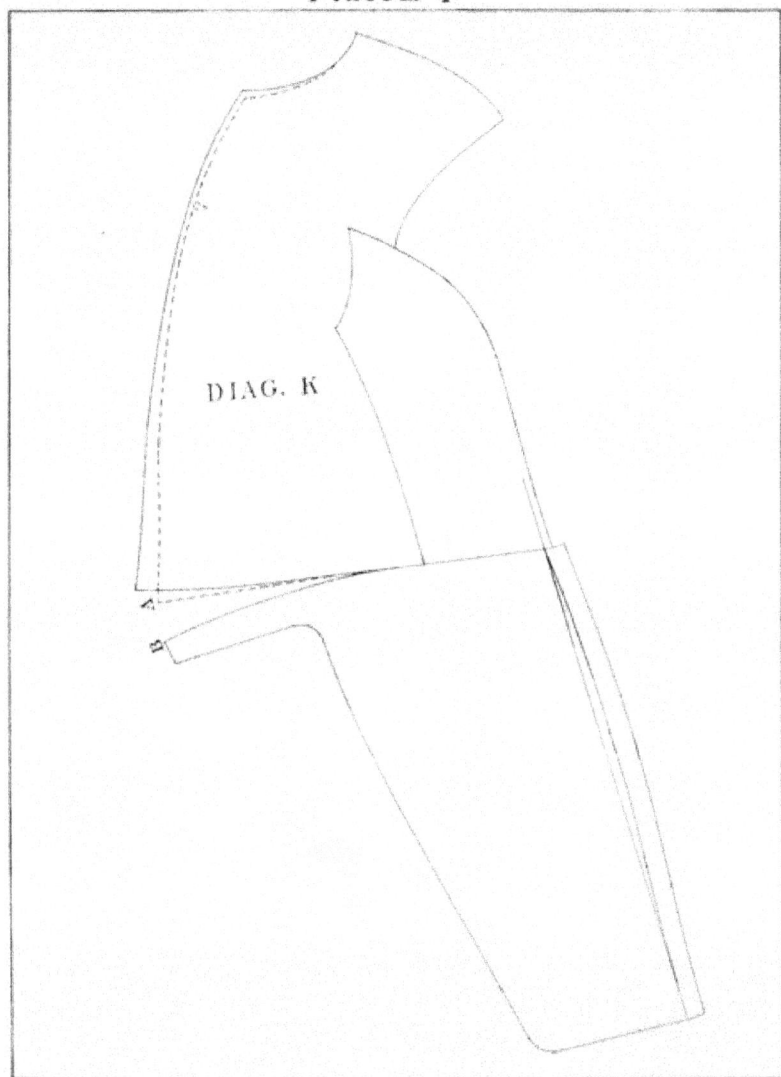

DIAG. K

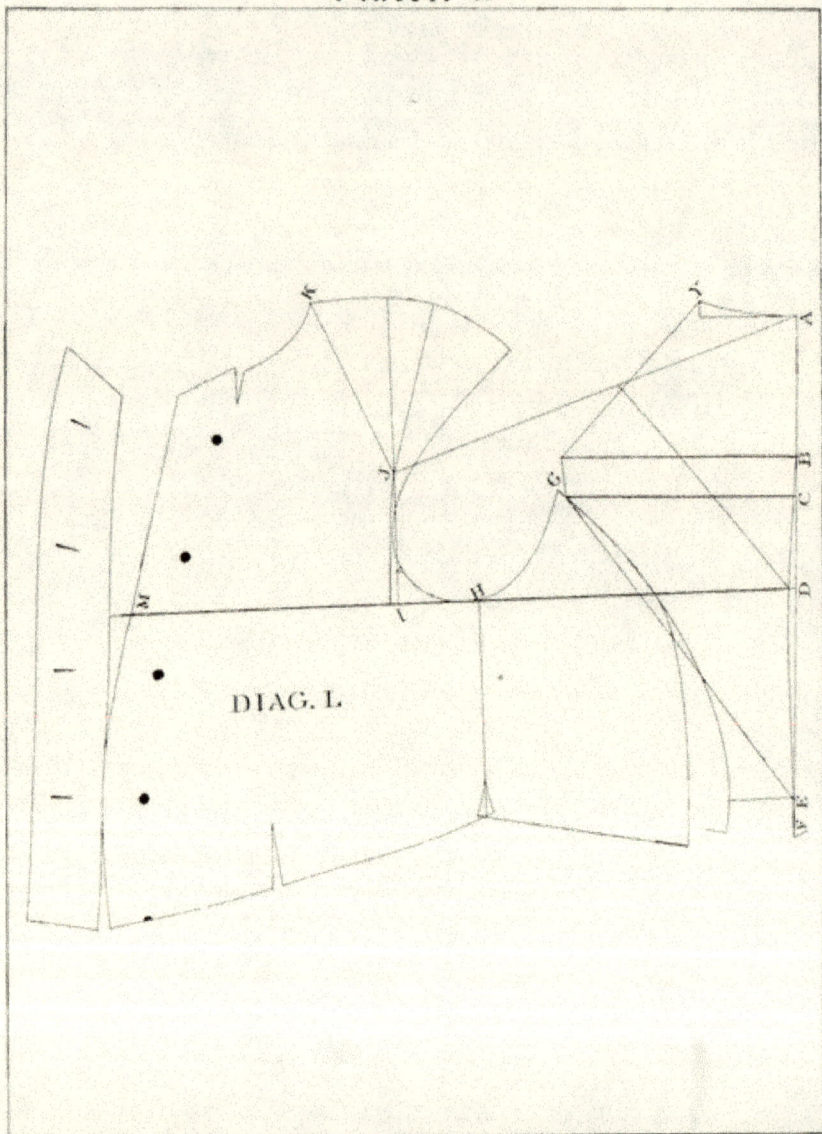

PLATE 8

DIAG. L

PLATE 8.

DIAGRAM L.

Showing a draft for a short, stout man. Drafted to the following measures: First down the back 9¾, 17, 18½; width of back, 8½; to the elbow, 20; length of sleeve, 30; scye, 20; length of shoulder, 14½; over the shoulder to the first measure down the back, 20½; short blade, 13½; breast, 43; waist, 44½; hips, 47; seat, 45; short neck and high shoulders. The lapel is cut a little hollow, from the top to the third button, and then rounded a little to the bottom; the latter is done to obviate the necessity of drawing the front edge in when making, and prevents a looseness on the edge when the coat is buttoned. The waist-seam coming below the most prominent round, we take a cut out at the bottom of forepart, to prevent looseness on the seam; in all other respects drafted the same as Diagram A. The seat being smaller than the hips, the top of skirt will have to be well rounded, otherwise the skirt will have too much spring and will flare out at the bottom.

PLATE 9.

DIAGRAM M.

Showing a draft for a double-breasted sack. First draw a line, A to F; move the square in ½ an inch at E. Then square across at A, B, C, and D, by the inside line. D to H, A. Square down from H by the outside line for the width of back at bottom. The upper part is drafted the same as Diagram A, except point K is carried ⅔ of an inch forward; N to O, ½; ¼ of an inch less from D to M, because of there being no seam off in front. In all other respects the upper part is drafted the same as Diagram A. S is the construction line, R is ½ an inch added to give ease when the coat is buttoned, and rounded out from S to R. Commencing at M, as shown in the diagram, square down from the point where the back touches the forepart at the round of side-seam to T, the seat line. T to C 2 inches. Now turn the back over on the forepart, touching at the lower part of the hollow at L, and on point C. Then mark the side of the forepart by that of the back. The forepart should be hollowed from the back at L 1 inch. For a close-fitting sack, the back should be hollowed more in the centre, and a cut taken out under the arm, from H down to the pocket. The average distance of the pockets from the bottom of the seye is 13 inches for men of medium height. Sweep for length in front as shown in Diagram A. The remainder of the draft as shown in the diagram.

PLATE 9

DIAG. M

PLATE 10

DIAG. N

PLATE 10.

DIAGRAM N.

Showing a draft for a single-breasted cutaway sack, drafted the same as Diagram M, allowing ⅛ of an inch at M for the button stand. Raise the front of gorge for all single-breasted sacks.

PLATE 11.

DIAGRAM O.

Showing a draft for a straight sack with the corners rounded in front. This is the same as Diagram M, allowing ⅝ of an inch for the button stand, etc.

PLATE 11

DIAG. O

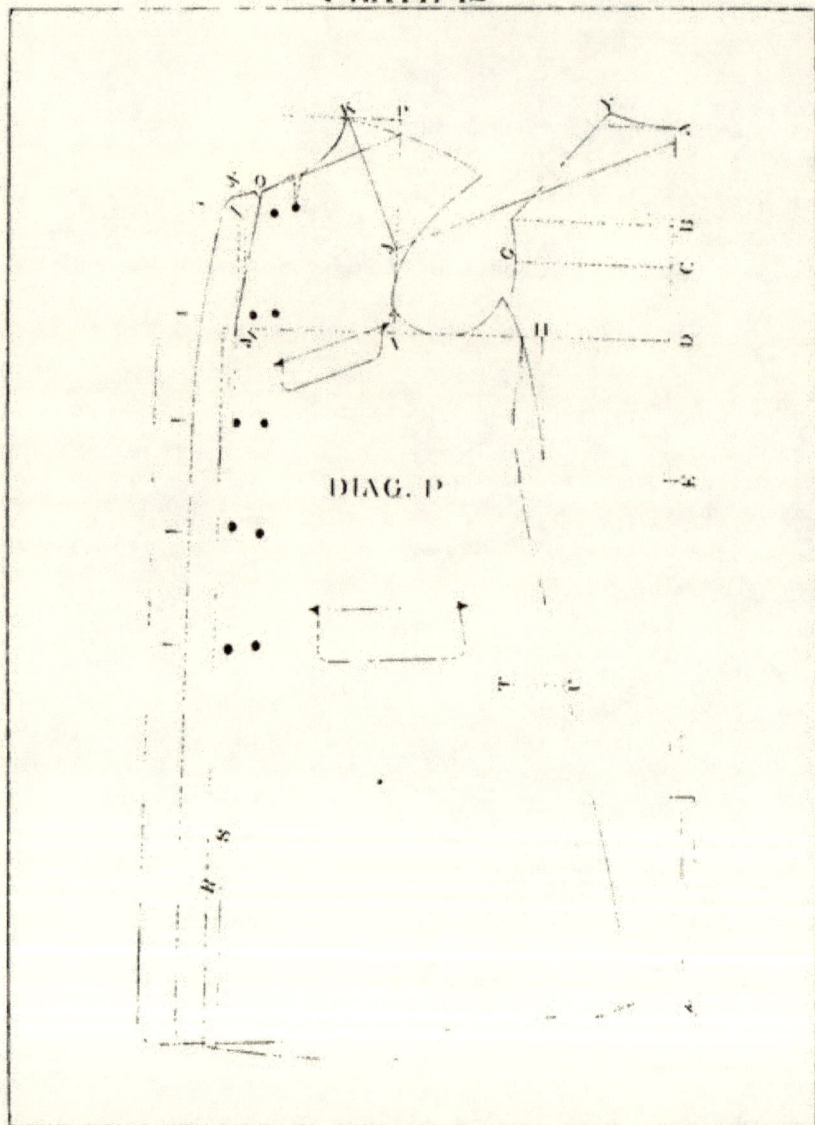

PLATE 12

DIAG. P

PLATE 12.

DIAGRAM P.

Showing a draft for single and double breasted oversacks. Drafted the same as Diagram M, except the additions to the measures. Sufficient must be allowed to go over the undercoat. In this draft the square is dropped one inch below A, to get point K. As a usual thing we make the following additions to the measures: From A to D, 1 inch; D to L, 1 inch; 1 inch to the length at shoulder, and 1 inch to the measure over the shoulder. Apply this measure to a point 1 inch above D. From T to U, 3 inches. This draft to be drafted from the same measures as Diagram A. Further comments will be found under the head of "General Remarks."

PLATE 13.

DIAGRAMS Q AND R.

Showing how a Glengary can be cut by an ordinary oversack pattern. Make the alterations as shown in diagram. By observing the back, it will be seen that the piece represented by the dotted line is cut off. When this is done add to the width of the back, as shown by the solid line, starting from nothing at top, ¾ an inch at the point where the mark is made to divide the piece cut off from the back and then running gradually out to 3½ inches at the bottom. The piece that has been cut from the back must be cut in two, the upper half to the forepart shoulder and the lower half to the top of side. Add about 1 inch to the blade and run it out at the bottom 2½ inches or more, if the garment is to be cut very full. In making up, the notches in the pieces cut from the back should be joined together for a sleeved garment; if without sleeves, cut by the circular line running from the shoulder to the side. Measure for the length of cape from the collar-seam over the shoulder to the ends of the fingers, and cut it as represented in the diagram. To have the garment go together fair mark the place for the notches on your sack pattern before marking out the Glengary, and have the notches come together in making. Measure for the notch in the side of cape by the back. If this is done carefully the garment will be properly balanced.

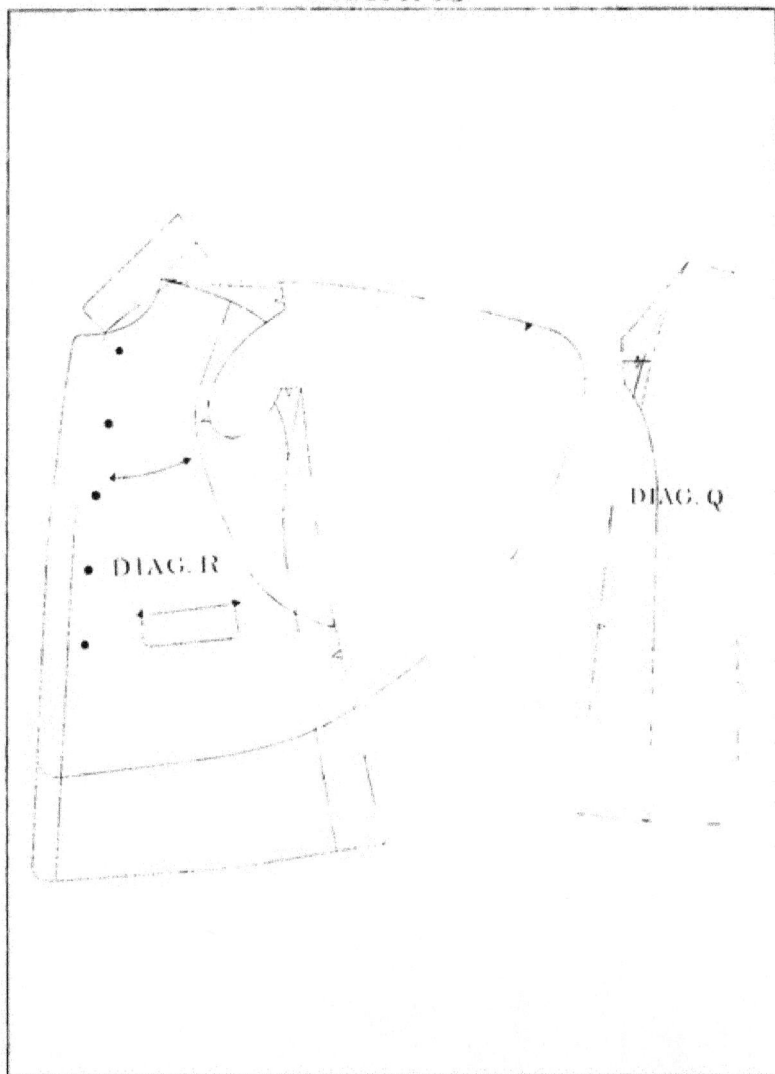

PLATE 13

DIAG. R

DIAG. Q

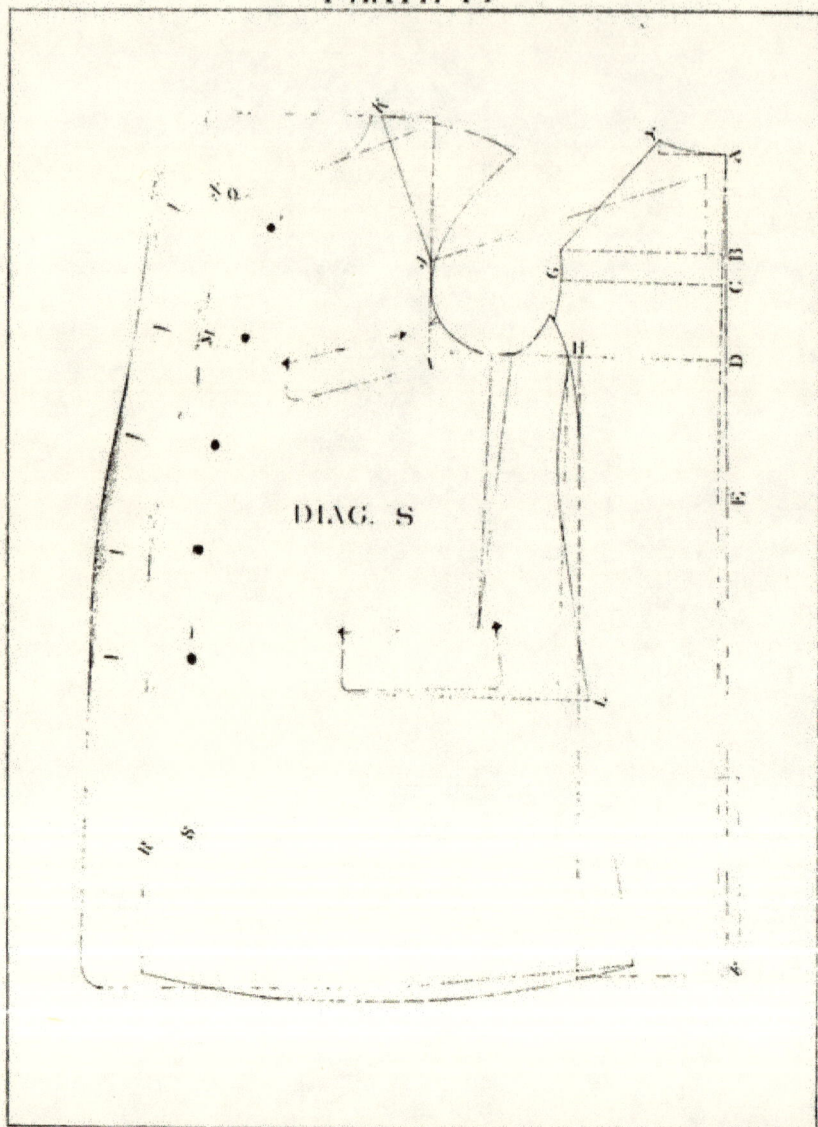

PLATE 14

DIAG. S

PLATE 14.

DIAGRAM 8.

Showing a draft for a sack overcoat, drafted by the same measure as Diagram L. The hip measure being 4½ inches larger than the breast, we proceed in the following manner to distribute the cloth. To make up for the difference between the breast and hips, we add 2 inches in front of the construction line, and add 1 inch to the blade measure. When the pattern is made, take away the inch that has been added to the blade, by a cut under the arm, running down to the pocket. To get point K, make a line 1 inch from the centre line of back between A and B, and mark a point 1 inch below A, and from this point square up from J to K. Apply the seat measure from line E back to L, including the back. The seat measure over the undercoat being 47 inches, to this we add 7 inches for ease, 3½ inches on each side. In all other respects drafted the same as Diagram M.

PLATE 15.

DIAGRAM T.

Showing a draft for a sack overcoat for a stooping figure, and drafted from the same measures as Diagram B. It will be observed that line S, the construction line, is brought back 1 inch at the waist, or say at a point 8½ inches below M (see diagram). Then place the square on this point, touching at M, and draw the line as represented. In all other respects drafted as Diagram P, except the line at the top of back is taken forward as in Diagram B, to give the effect as there noted. The object of changing the construction line is to bring the cloth to the front at bottom, and to reduce it a like amount at the back. We prefer to do it in this way, rather than to force the cloth forward by straightening the shoulder—that is, carrying point K forward. For a very erect figure we carry the construction line in front of line R, or just the reverse of this diagram. By this means it takes the cloth from the front at bottom, and takes it back, thereby giving more spring, which is required for the erect figure. The construction line must be moved back or forward according to attitude. This must be left to the judgment of the cutter as to how much to deviate.

PLATE 15

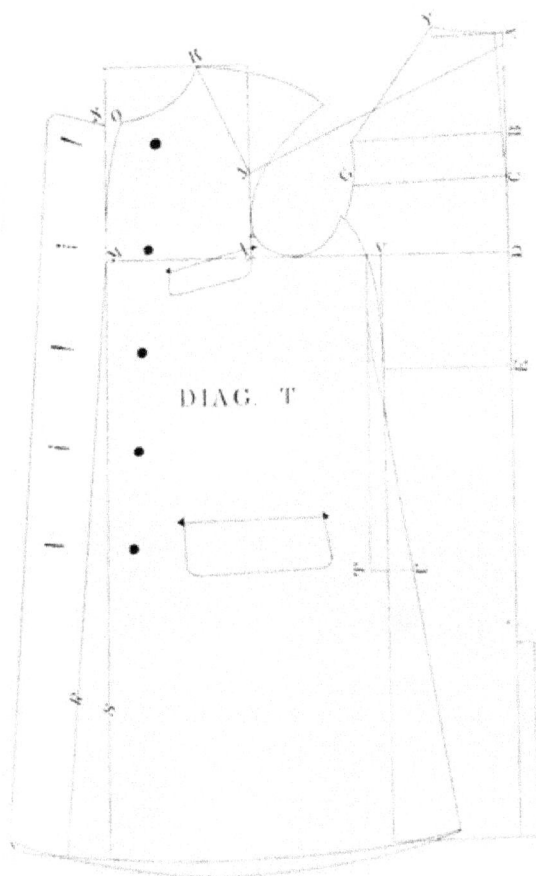

DIAG. T

PLATE 16

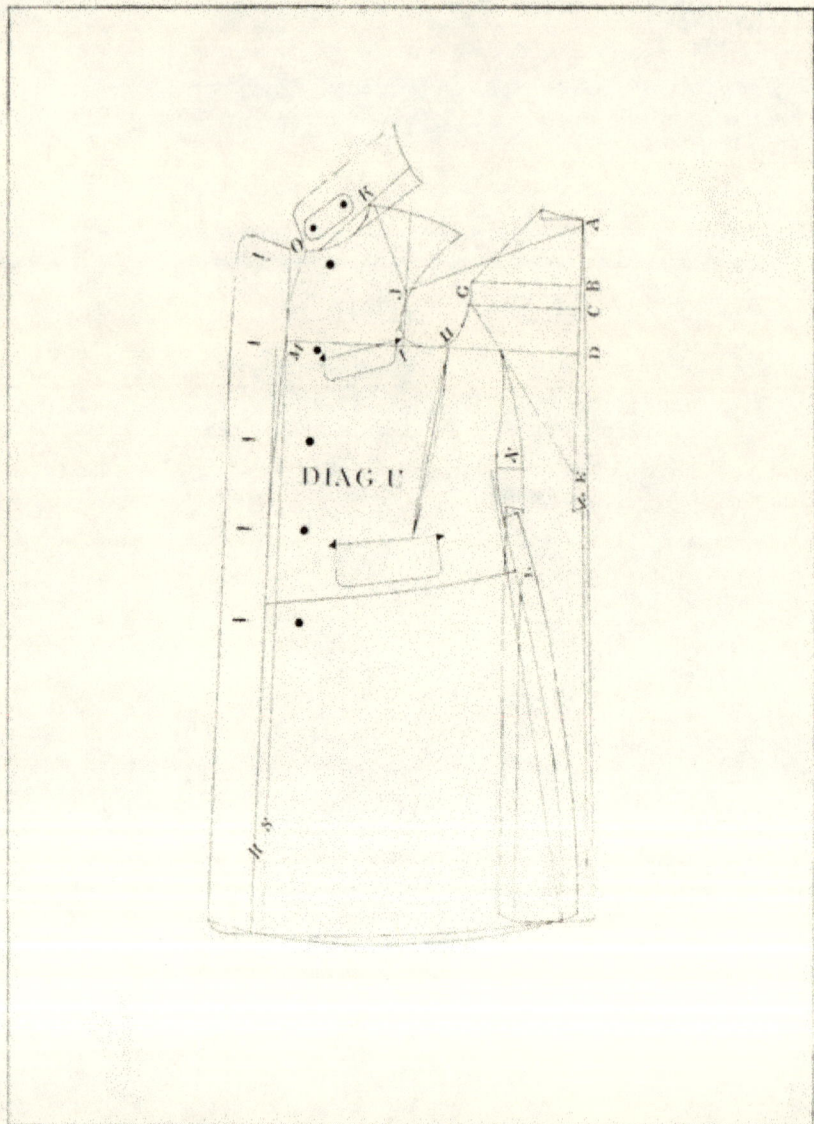

DIAG U

PLATE 16.

DIAGRAM X.

Showing how to produce a surtout ulster or ulsterette, drafted to the same measure as Diagram A, with the additions to the measures as has been previously described. From line R over to L, mark the seat measure. Hollow the forepart 1 inch from the back. Mark the sideseam of forepart as represented, and draw the plait line, as shown. Round out from the line at L ½ an inch and run it in at the bottom of skirt. The cut under the arm can be taken out or not. If it be desired to have the garment fit close, take it out as represented. The opening at O is ¼. The width of back at waist to fancy.

PLATE 17.

DIAGRAM V.

Showing how a cape or cloak can be cut by an ordinary coat pattern. First lay the back in the double edge of the cloth, then place the forepart with the shoulder point touching the back as represented in the diagram. This position will produce a ½ circle. Take the cut out at E, as shown by the dotted lines. If it be desired to increase the circle, close the back and forepart nearer together at the shoulder; if entirely closed at the neck it will produce a ¾ circle. For a cloak all that is required is to increase the length. For an ordinary cape the length would be, say from A to B 20 inches; E to C over the shoulder 21½ inches; and from the front of neck to the bottom in front to D 18½ inches. When cut in this way it will hang even. The collar should be cut the same as Diagram R. Plate 13, or if to button on the coat put a band around the neck wide enough for button-holes except in front. The front button-hole should be in the seam, and the band cut to a point in front to prevent it from crowding the collar up.

26

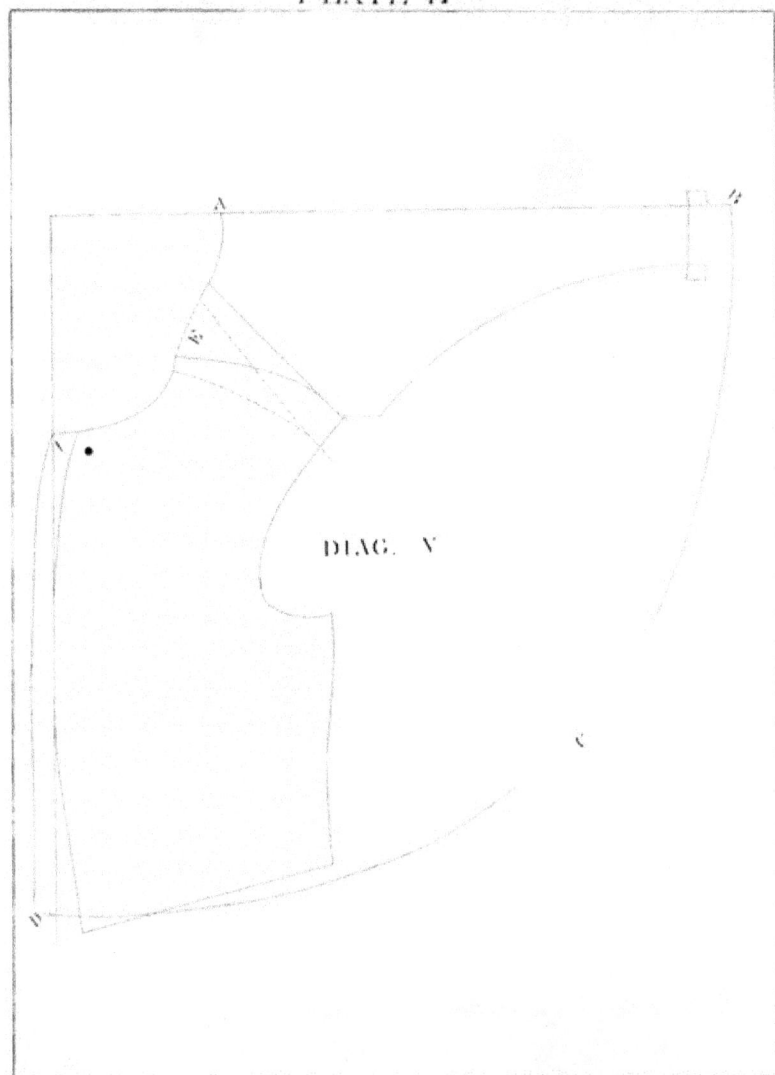

PLATE 17

DIAG. V

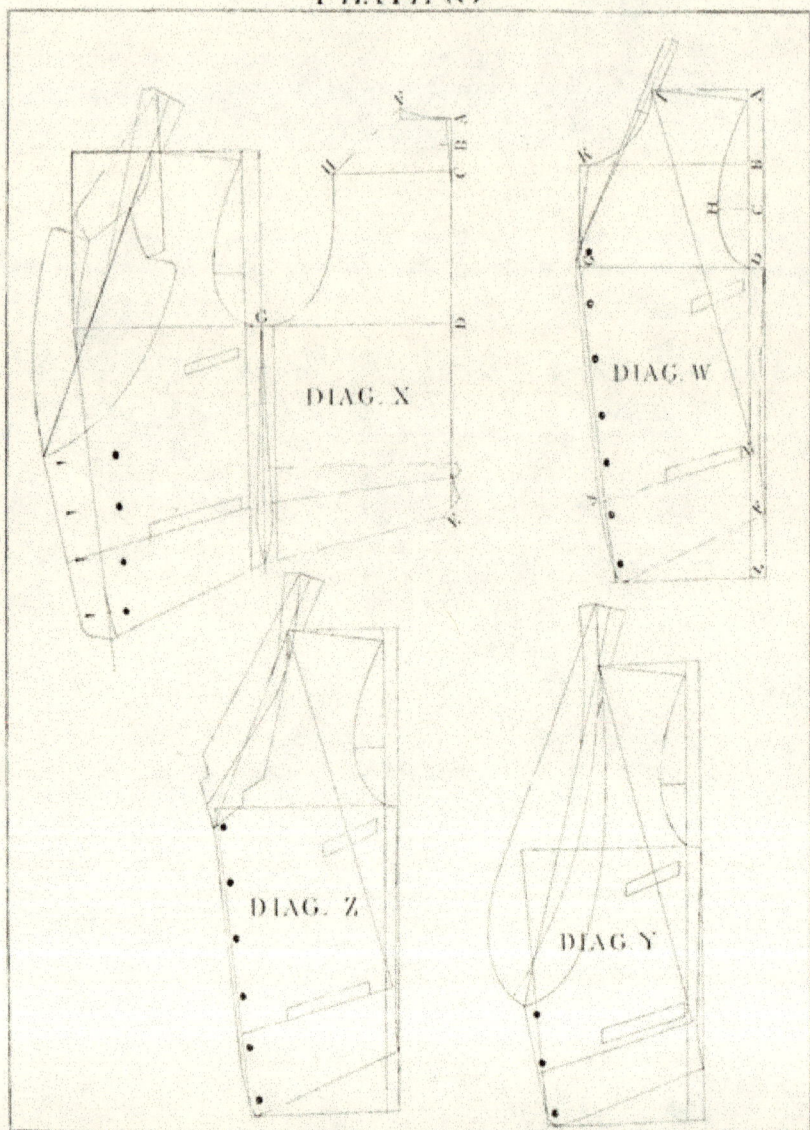

PLATE 18

DIAG. X

DIAG. W

DIAG. Z

DIAG. Y

PLATE 18.

DIAGRAMS W, X, Y and Z

Showing how to draft different styles of vests.

DIAGRAM W, THE STANDARD FOR ALL SHAPES.

To form the draft: First draw a line down the edge of paper or cloth; make a second line inside and one inch from the first; this we call the construction line. Mark on this line from A to B ½ of the breast measure less ¾ of an inch from A to D, ½ breast measure, or if cut to measure apply the shoulder length from I down to the breast line or bottom of the scye. From D up to C ½; C to B ⅟₂; A to I ½ of an inch more than ¼ of the breast measure. Measure down from I to E 19½ inches to the top of the hip, or 2½ inches more than the natural waist measure down the back. Make a point here half way between the first and second line; put the corner of the square on this point with the long arm resting on I, and square over to J. Measure from the inside line at E over to the first line back of J, ½ the waist measure. Then put the square on this point and draw a line up to G, G being ½ the breast measure from D. Square up from G to K; take ½ an inch off at this point and round down to G. Curve the neck as shown in the diagram. Measure the length of front, and from this point square over to L; L to F ½; from this point mark the bottom line of forepart. If the draft is intended for a double-breasted vest with the lapels cut off, cut as the first line front of the button marks; if for a single-breasted, add the inturn as represented; if for a double-breasted vest with the lapels cut on, mark off, say ½ of an inch all the way down the front. Now finish the draft as shown in the diagram.

DIAGRAM X.

Showing how to form the back. First draw a line from A to E; then mark from A to B 1½ inches; A to C ¼ of the breast measure; B to D ½ the breast measure; or if cut to measure, take the measure down the back as taken for the coat. Square across at A, B, C and D; D to G ¼ the breast measure and 1 inch more. Measure the length of the forepart shoulder-seam, and apply this length from F to H on line C for the length of the back shoulder-seam; the width of back at the waist ½ the waist measure and 1 inch more. Now finish the back as represented in diagram. The forepart adjoining this back is for a double-breasted vest with the lapels cut on.

DIAGRAM Y.

Showing a draft for a dress vest. The front line is carried full ½ an inch forward at the top button mark. In all other respects drafted the same as Diagram W, except opening low in front.

DIAGRAM Z.

Drafted the same as Diagram W, the only difference being in the collar.

27

PLATE 19.

DIAGRAMS AA, BB, AND CC, SHOWING HOW TO DRAFT PANTALOONS.

AA. THE FOREPART.

First draw a line A to D, the outside length, D to B the inside length, B to C ¼ of the outside length, D to L ½, of the seat measure, increased to an ¾ or more for large sizes. Make a second line from B to L. Square across by the outside line at A, B, C and D; B to E ½ less ⅛; D to E at the bottom the same. Draw the centre line E, E; E to F 1½ inches; E to G ¾; G to K½; G to J ¼; from K to H ½. Square up from G to the top at O, draw a second line from H to O, making the distance between the two lines at top half of H G; G to N ¼; curve from N to J, and from N to K as represented in the diagram. L to E is half the bottom width; E to N the second half. Draw a line from J to N, and from K as represented; O to M ½ an inch less than ½ the waist measure; that is, ¼ of 15¾, 31½ being the full waist, all the other divisions are divisions of the ¼ seat measure 19, the full seat measure being 38. Round down from M to B. Finish the bottom of the forepart as represented in the diagram.

DIAGRAM BB.

Now place the forepart as represented in Diagram BB. Mark point I half way between H and J; continue line F up to C ¾ above the top of front; draw the seat line from I to C; from I to P ½; then curve as shown in the diagram. Make the width of knee and bottom to the measure. Sweep from M to D, then measure from a point half way between the lines at O to M, and from C to D ½ the full waist measure and one inch more; add 2 inches and the seams to the seat measure; take the inch that has been added to the waist out at N; hollow the top of the seat as shown in diagram, and the bottom also finished as shown. For tight-fitting pants reduce the fork point of the underside, and take an equal amount off all the way to the bottom; this prevents the inside seam from becoming too hollow. We think the straighter the inside seam is cut the better the pants will fit.

DIAGRAM CC.

Showing the changes to be made for stout waists. We invariably raise the front above the square line ¼ of an inch for every 2 inches that the waist increases in size, starting the additions from 35. Draw line H, square up with G, and round the lines G and H as shown. The distance from I to P should be reduced, for it usually occurs that men with large waists and seats have small thighs in proportion to the other measures. If the seat measure is smaller than the waist, it will not be necessary to take the cut out at N as in Diagram BB; in

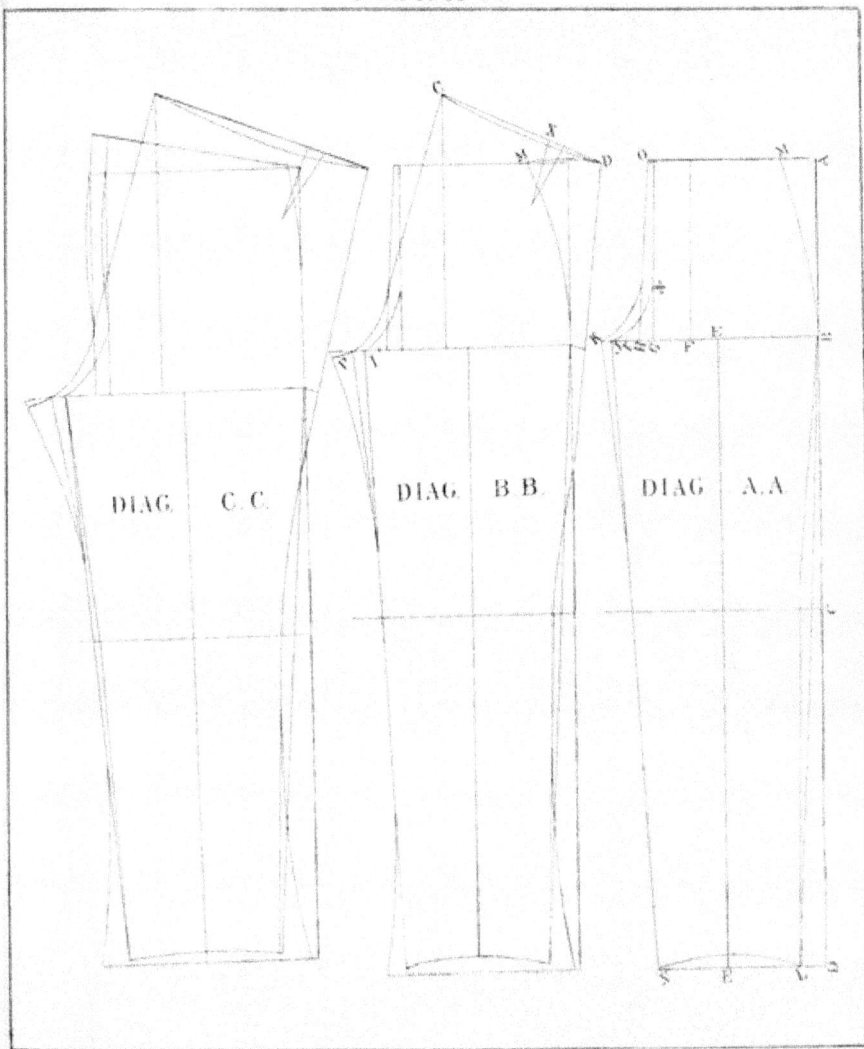

PLATE 19

DIAG. C.C.

DIAG. B.B.

DIAG. A.A.

PLATE 20

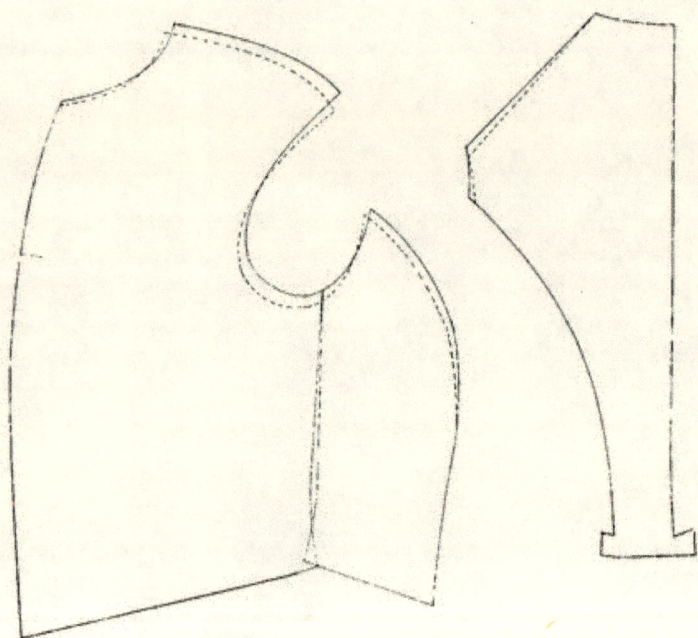

reducing the size of the knee it should be done at the outside seam. If it be desired to cut spring-bottom pants, the line at L should be taken further in, and also the inside line at X, it being necessary to have a narrow forepart so that the sides can be stretched down and shrunk in at the centre better, for it is almost impossible to shrink a wide forepart in the centre sufficient to produce the desired effect. What the forepart has been reduced in width must be added to the back part. Line L should be dropped ½ an inch for the extra length required for spring-bottom pants; it should also be rounded on the front. Pantaloons cut for men with large calves and set back should have a round cut on the underside below the knee at the outside seam, and fulled on to the forepart; this will require to have an additional length also, more or less length according to the size of calf, and the fulness pressed back to the centre. It is also a good plan to shrink the underpart above the knee, to take the loose cloth from this part and make it conform to the shape of the leg.

For men with knees turned out the inside seam should be more hollowed; if the knees are turned in then hollow the outside seam more. For men with very large seats, requiring greater length to go over this part, point F should be brought nearer to E, and point C raised more above the square line of front. If for small seats point F should be carried more forward. Point C remains the same as in Diagram BB; how much these points should be moved must be left to the judgment of the cutter, for no special rule can be laid down to govern these changes, the object being to give the cloth where it is required.

Measures for pantaloons.—When using the measures observe the following rule: add a seam to the rise, 2 inches and the seams to the seat measure—we have added 3 inches to the close thigh measure; add the seams in addition, also add the seams at the knee and bottom. Measure the thigh measure 1 inch below the crotch line on the non-dress side. The dress not taken out below 32, and seldom above that for stock pantaloons.

Rise......	7½	7¾	8	8¼	8½	8½	8¾	9	9	9¼	9½	9¾	10	10¼	10½	10¾	11	11¼	11½	11¾	12	12¼	12½
Leg......	24	25	26	27	28	29	30	31	32	32½	33	33½	33¾	34	34	34	33½	33½	33	32½	32	31½	31½
Waist......	24¾	25¼	25½	26½	27	28	28½	29½	30	30½	31½	32½	34	35	36½	38	39½	41½	43	44	45½	47	49
Hip......	28	29	30	31	32	33	34	35	36	37	38	39	40	41	42	43	44	45	46	47	48	49½	50½
Thigh......	19½	20	20½	21	21½	22	22½	23	23½	24	24½	25	25½	26	26½	27	27½	28	28½	29	29½	30½	31½
Knee......	14½	14¾	15¼	15½	16	16½	16¾	17	17¼	17½	17¾	18	18¼	18½	18¾	19	19	19¼	19½	19½	19¾	20	20½
Bottom......	14¾	15	15½	16	16½	16¾	17	17¼	17½	17¾	18	18¼	18½	18¾	19	19¼	19½	19½	19½	19¾	19¾	19¾	20

The two opposite tables of measures are designed for the purpose of making block or stock patterns. The chief object being to instruct the student by having the measures to refer to, he can select any measure from either of the tables, and can draft a pattern of any size desired from 24 to 48 inches, breast measure; which we believe will be instructive to a new beginner. When drafting the small sizes add ¾ of an inch to the measure of 24, 25, and 26; ½ an inch to 27, 28, 29, and 30; ¼ of an inch to 31, 32, 33, and 34. Reduce the large sizes as follows: 41, 42, and 43, ¼ of an inch on each size; 44, 45, and 46, ½ an inch on each; 47 and 48, ¾ of an inch. The small sizes require to be cut larger in proportion to medium sizes, *because there is not cloth enough to stretch to make them easy*, the large sizes having too much we reduce them as directed. To find the seat line, add the length of natural waist from the bottom of the seye to the rise of pantaloons, and take off 1½ inches. Take, say size 36 breast, from the bottom of the seye to the waist line, 8 inches; the rise of pantaloons, 38; hip measure, 9½; rise, 8 and 9½, 17½ inches; 1½ inches off makes the seat line 16 inches from the bottom of seye. For sizes above 42 breast take off 2 inches. When drafting the small sizes, from 24 to 29, drop the back ¼ of an inch below P instead of ½ an inch as in diagram A; from 30 to 34, ⅜ of an inch; 35 to 40, ½ an inch; 41 to 46, ⅝ of an inch; 47 and 48, ¾ of an inch; or use the measures to get the height of shoulder, the seat measure use to get the spring of skirt.

We never take a balance measure to the natural waist, believing if the measures for the upper part of coat are correct the lower part will fall into its proper place. The tables of measures we have estimated from a very large number taken by us during the past 30 years, and if used as directed will produce easy garments.

SACKS FROM 24 TO 34 INCHES.

	24	25	26	27	28	29	30	31	32	33	34
Breast	24	25	26	27	28	29	30	31	32	33	34
Depth of seye on the back	6¼	6¾	6½	6¾	7¼	7½	7¾	7¾	8	8¼	8¾
Natural waist	11¼	11¾	12½	13	13½	14¼	14½	15	15½	16	16¼
Length	19	20	21¼	22¾	23½	24¾	25¼	26¾	27½	28	28¼
Width of back	5¼	5½	5½	5½	6	6¼	6¼	6¼	6¼	6¼	6¼
To elbow	12	12¾	13¼	14	14¼	15½	16¼	16¾	17½	18¼	19
Sleeve length	20	21	22	23	24¼	25¼	26¾	28	29	29½	30
Around the seye	12	12½	12¾	13¼	13½	13½	14¼	15½	15½	15¼	16¼
Shoulder length	9¼	9½	9¾	10¼	10½	10½	10¾	11¼	11¼	12	12¼
" height	12¾	13¼	13¾	13½	14¼	14½	15	15½	15½	16¼	16½
Short blade	7½	8½	8½	8½	8½	9¼	9½	9½	10½	10¼	10½
Breast	24	25	26	27	28	29	30	31	32	33	34
Waist	24	24½	25	25½	26	26½	27½	28½	29	29½	30½
Hip	25¼	26¼	26¼	27¼	27½	28½	29	30½	31	31¼	32¼
Seat	26	27	28	29	30	31	32	33	34	35	36

FROCKS FROM 34 TO 48 INCHES.

	34	35	36	37	38	39	40	41	42	43	44	45	46	47	48
Breast	34	35	36	37	38	39	40	41	42	43	44	45	46	47	48
Depth of seye on the back	8½	8½	8¾	9	9¼	9½	9¾	9½	10	10½	10½	10¾	10½	10¾	10¾
Natural waist	16½	16½	17	17½	18	18½	18½	18½	18½	18½	18½	18½	18	18	18
Full length of waist	17½	18	18½	18¾	19½	19½	19½	19½	20	20	20½	20½	20½	20½	20
Full length of coat	35½	36½	37	38	39	39½	40	41	41	41	41	41½	41	40½	40
Width of back	6½	7	7½	7¼	7½	7¼	7½	8	8½	8½	8¼	8½	8½	8½	9½
To elbow	19	19¾	20	20½	21	21½	21½	21½	21½	21¼	21¼	21½	21½	22	22
Sleeve length	30	30½	31	31½	32	32½	33	33½	33½	33½	33¾	33½	33½	33½	33½
Around the seye	16½	16½	17½	17½	18	18½	18½	19	19¼	19½	19½	20	20½	20½	21½
Shoulder length	12½	12½	12¾	13½	13½	13½	14	14½	14½	14½	15	15½	15½	17¾	15½
" height	16¼	17½	17¾	18½	18½	19¼	19½	20	20¼	20¼	21¼	21¼	21¾	22	22
Short blade	10½	11¼	11¼	11¼	11¾	12½	12¾	13¼	13½	13½	14½	13½	14½	14½	14½
Breast	34	35	36	37	38	39	40	41	42	43	44	45	46	47	48
Waist	30¼	31	32	33	34½	35½	37	38½	40¼	42	43½	44½	46	47½	50
Hip	32¼	33	34	35	36½	37½	39	40½	42½	44¼	45½	46½	48½	50	53
Seat	36	37	38	39	40	41	42	43	44	45	46	47	48	49½	50½

31